美好家居

图书在版编目(CIP)数据

美好家居／(以)奥丽·罗宾逊编著；齐梦涵译．—桂林：广西师范大学出版社，2018.5
ISBN 978-7-5598-0726-7

Ⅰ．①美… Ⅱ．①奥… ②齐… Ⅲ．①住宅－室内装饰设计 Ⅳ．①TU241

中国版本图书馆 CIP 数据核字(2018)第 052111 号

出 品 人：刘广汉
责任编辑：肖　莉
助理编辑：齐梦涵
版式设计：吴　迪
广西师范大学出版社出版发行
（广西桂林市五里店路9号　邮政编码：541004
　网址：http://www.bbtpress.com）
出版人：张艺兵
全国新华书店经销
销售热线：021-65200318　021-31260822-898
广州市番禺艺彩印刷联合有限公司印刷
（广州市番禺区石基镇小龙村　邮政编码：511450）
开本：889mm×1 068mm　1/16
印张：18　　　　　　字数：30 千字
2018 年 5 月第 1 版　2018 年 5 月第 1 次印刷
定价：188.00 元

如发现印装质量问题，影响阅读，请与印刷单位联系调换。

美好家居

[以] 奥丽·罗宾逊 编著

齐梦涵 译

广西师范大学出版社
· 桂林 ·

images Publishing

我一直认为自己就是一个家，
我把我的整个人生都带进家中。
无论家是什么，我都是它。
本书献给我深爱的孩子们，
他们是家庭对于我来说意味的一切。

目录

8 家：这是你倾注一生的地方
11 设计的力量

14 认识到你属于这里
31 一个独特的、私人的空间
36 阳台上的冒险
40 起居室
49 用内在意识设计你的家
70 一餐不仅包括食物
76 光之厨房
89 呼吸新鲜空气
97 是什么让我们在餐后留在桌边？
120 灵魂需要空间
123 什么样的光线最适合你？
128 有时勇气要求敢做
139 幽默、色彩与欢乐
146 我们都需要安静的时光
168 纺织品
176 家庭思维
189 一座宁静祥和的岛屿
195 一个房间的感觉
223 居家办公
238 仿若漫步田野
254 绿色假期

282 设计师与建筑师
283 图片版权信息

家

这是你倾注一生的地方

你应该知道那种你到了一个特别令人愉快的地方之后不想离开的感觉。即使在你离开以后，你仍然保有一份深刻而难忘的回忆，还有对再次前往那里的热切期盼，你渴望在那里呼吸、放松、休息、安躺；去嗅闻、品尝、与触摸。

你试着去了解，这是什么地方？为何你会安睡得如此深沉？为何食物如此美味，沐浴如此令人愉悦放松？这是因为风景，还是因为房子的布局？它在讲述的是一个途径还是一个故事？透过窗户能看见远方的景色吗？你的家能成为一个这样的——舒适、安全、放松，为呼吸、饮食、拥抱和安慰而存在的——地方吗？你是否也想生活在一个滋养你灵魂的健康、快乐、充满活力的地方？你是否也想体验这种把家当作避难所和自由之地的感觉？

本书将展开一段探索这样的家园的旅途，我们在那里将获得极大的满足，除了家能提供给我们的东西以外再无须其他，我们从那里到任何地方都只需骑自行车即可抵达。在那里，阳光能够照射任何角落，太阳不再暴晒，而是轻抚。那里微风习习，树木献出香甜的水果，植物和草药在花圃中生长。家是生活的中心，是人们工作、休闲和睡觉的地方，是人们真正度过人生的地方。

我们的旅程花了两年的时间。我们遵循着"绿色家园"的原则，把环境纳入考量，并将其变成对于家园和生活在其中的人们来说不可或缺的一部分。我们遇到了为构建可持续发展而努力着的建筑师，拜访了一些整合了环境元素的家庭，还发现了建筑保护性住房及使用环保或当地材料以创新技术或传统工艺改建的住房。从这些住房中我们发现，建筑师们对房屋的选址和该地的光照、风向、开口、隔热、花园、周边环境和景色等属性看得同样重要。

但是更重要的是，我们发现了一些寻求温和的生活方式的人们，他们消耗得更少，旅行得更少，浪费得也更少，他们生活在家中，生活在周边的环境和社区中。

他们是具有环境、社会和经济意识的人，他们努力只使用必要的资源：明智地消费、保存和享受家居环境及其潜能。我们遇到过到从来不在超市拿多余塑料袋的人，他们喜欢在家吃饭，而且积极进行垃圾分类。

在这次旅行中，我们也发现了"绿色家园"与"情感家园"之间的联系。在一些住宅中，我们发现了居住者对环境的细致关注，还有他们对土地、对社区、对宇宙、对他人的特别关照。我们发现了一些住宅中体现出的情绪上的精确性，这其中体现出居住者知道如何区分哪些事情是重要的，哪些事情是无关紧要的。

我相信家是一个你应该去爱，去生活的地方。

我相信家应该包含一个人整个一生的地方，它是一切的起点，是你出发的地方，也是你返回的地方。它代表了我们是怎样的人，以及我们想成为怎样的人。

在我看来，一个家应该是我们的房间、餐厅、社交俱乐部和创意工作室，如果可能的话，这里也应该是我们工作的地方。

一个家是一个我们可以一次又一次糅合的地方，它是巢穴形式的爱与关怀。

我个人觉得做家务是对生活充满爱意的表现，我喜欢为我爱的人们做饭，安排房间，触摸家中的每一件物品，去安排，去整理——做这些事是在用另一种方式对家人诉说："我在这儿陪着你，照顾着你，我爱你，我在为你创造一个家。"

我们并不都是在舒适和关爱的环境中长大的，我们也并不都生活在一个美好的环境之中，但是我们有权为我们自己寻找一个这样的特殊场所。诚然，设计本身并不会带来内在的快乐，但是平和的外部环境对调节心情是十分有帮助的，也有利于创造一个能够包容和保护家庭成员的空间。在我看来，绿色家园是一个充满可能性的家园。它为家庭提供了力量和支持，是我们开展所有日常活动的场所，也是我们生活的中心环节。

我从自己多年来拜访别人的家的大量经验中，懂得了一个完美住宅并不只是为那些仿佛从广告中走出来的完美家庭而存在的道理。它无差别地为一切灵魂、家庭提供一个舒适的地方，无论那个灵魂或者那个家庭是什么样子的。

本书为那些希望给家庭带来安静、祥和和绿色的人们编写，为那些思考和憧憬可持续发展和生态平衡的人编写，就算他们已经踏上了这个旅程，就像我一样，这本书对他们来说依然是有益处的。

本书的核心是家是家人的情感归属之场所，凭借着这个核心，本书旨在鼓励人们于生活处着手，加强情感的家园在我们生活中的意义。

我邀请您来和我们一起旅行，请您写信给我们，或在我们的网站及Facebook上留言，分享您的经验。我希望我们所有人都能获得一个我们可以在那里健康生活的、充满欢乐和创造力的家，一个我们可以实实在在地生活于其中的家，一个真正的家。

此致敬礼

奥丽·罗宾逊

设计的力量

xxxxx x xxxxx

家居装饰在本质上是一种特权,而不是生活必需品。毫无疑问,生活中有许多更要紧的优先事项,比如健康、营养、房租、教育、衣着,等等。任何与设计有关的支出,都是与更新居住设置有关的支出,因此从表面上看来,它或许是不必要的。这样的支出总是会以牺牲其他东西为代价,被牺牲的也许是娱乐预算,也可能是我们的银行存款。近几年来,我们越发意识到设计的力量能够影响我们的情感、心理和社会福祉。设计已经成为我们生活之中不可或缺的一部分,它甚至也是我们愿望中的一部分了。现在,全世界的人们都能通过社交网络,或通过手机接收到的源自应用程序、网站、杂志和图书的大量图片,注意到那些具有力量的设计。这种涌入强大,鼓舞人心,它呼吁我们采取行动。人们都希望过上更好的生活,无论这当中涉及到的是服装时尚还是家居设计。我相信设计的力量,也欣赏其优点。我能看到用合理的预算换来的轻质地毯、几个枕头和窗帘会怎样彻彻底底地改变房间的外观;只需简单地整理并用有香味的地板清洁剂清洁房屋,就能在何种程度上提升房屋的观感;洗衣、烹饪和鲜花的香味总能成功地创造出一个愉快的家居氛围。从这里开始,任何事情都是可能的了,我们都应进入这个我们能消费得起的设计世界。通过我们能够负担得起的设计,你可以在低预算的范围内做自己想做的事情:改装家具或是给家具上漆,创作个性艺术品,展示收藏品,摆放家人照片,展现你的创造力,并相信设计和健身一样,虽然在开始时需要一定程度的努力,但是一旦我们适应了,就会感觉好极了。你的眼睛、头脑和心灵越开放,你越会发现家中可以装饰的地方,也就有越多的机会展现在你的面前。一件设计突出的灯罩会成为起居室里的明星;一张改造过的旧扶手椅,也会给家带来新鲜的色彩;即使是一个简单的食物罐子,只要涂上颜色,也能成为一件漂亮的装饰品。设计的力量,记住并利用它。

RETURN HOME,

take a breath, feel what it represents for us, absorb the feeling of security, the atmosphere, values, authenticity, earth, and the site;

KNOW THAT YOU BELONG.
认识到你属于这里

The knowledge that the home will always be there for us when we return from a hard day's work or from a brief visit to the neighbor is comforting, happy, and beneficial knowledge. There is nothing like encountering a familiar place that allows us to realize our daily routines: throw down your bag, take off your shoes, and put down the packages. This knowledge is strengthened upon entering: warm and natural materials pique our memory of what has always been before us, and it has the power to strengthen our sense of belonging and intimacy. The preservation of this century-old Templar house used materials that were already part of it, out of respect and appreciation of the past, and promise and care for the future. The old plaster was renewed, the wood banister was recreated, and the old wooden doors were renovated.

How do we choose to design

OUR HOMES?
MORE IMPORTANTLY, HOW DO WE CHOOSE TO LIVE IN IT?

The concept of "home" has positive connotations of warmth and joy in our consciousness. If we are lucky, our home provides us with breathing space, comfort for the soul, a safe place to return to; but even if our big house does not provide optimal or perfect support, our hearts still beat with the hope of improvement, of creating a home that is connected to the environment, people, and the local soil. Design is not everything, of course, but it has the power to contribute to the senses, even subconsciously, creating a home: the use of local materials, poured concrete, wood for heating, plants, and textiles all strengthen the sense of "hominess" inside and outside. Environmentally considerate construction—the knowledge that violence towards the earth is not necessary and avoiding the use of heavy earth-moving machinery—is something conciliatory that acts on awareness. Here, the steps leading to the house are built on the stone of the hill, taking into account the land's natural contours, honoring the site that was here before and will remain after we are gone.

HOW DOES THE ENCOUNTER WITH A HOME OCCUR?

HOW MUCH DOES THE FOYER SYMBOLIZE WHO YOU ARE AND WHAT IS HAPPENING INSIDE?

A foyer that is well taken care of and nurtured is like an aromatic cake straight out of the oven: it is wholly an invitation.

Personal and aesthetic style are marvelous tools for sending an emotional message of care and love, of hugging and relief. Colored wooden boards, well-tended flowers, a renovated chest of drawers, and terrazzo tiles taken from an old house tell the story of the home and its residents—traders in aesthetics and handwork. The foyer is a wonderful opportunity to show who you are. Renovated furniture from Grandma's home or the flea market, fixed up with a loving hand and materials and paints that do not harm people or the environment; such a foyer makes it possible to breathe in the home, and experience it intimately. Each day it seems to say, "Welcome. It is nice of you to come."

AND HOW DO YOU WISH TO EXPERIENCE YOUR HOME?

Slowly? Intimately?
Perhaps have it exposed to the eyes of all?

The entrance path creates a gentle passageway between the street and the house. Walking down the path is a sort of invitation. We have few opportunities to create an unmediated, borderless connection between nature, the mountain, the woods, the house, and the courtyard. In such cases, one can utilize the generous gift provided by nature—a directly accessible garden that does not require construction or maintenance. The natural woodland provides an ecological, natural, and genuine garden.

Right from the entrance to the garden, one can see how the exterior design is carried out based on an integrated perception of the house and the garden. The passages between them become blurred and

ONE ORGANIC AND WHOLE UNIT IS CREATED.

THE ENTRANCE TO THE HOUSE

is a part of the garden's division into areas of hospitality, bathing, relaxation, cooking, and a bonfire, enabling the full range of life's activities to be experienced in the garden,

DIVIDED BY MAGICAL LUSH AREAS.

The form-shifting garden transforms colors, scents, and sights each season and creates an instant connection to nature, inviting us into a place of refuge and relaxation from the busyness of day-to-day life. Each step and every angle reveals another sight, a surprising blossom, a slightly different appearance. The seasons and water in the garden attract nesting birds and butterflies—a small slice of nature in the heart of a city.

THE HOME IS WAITING FOR YOUR RETURN,

ready to welcome you each day anew. Do you stop for a moment to prepare for this daily meeting? The front yard invites you to rest a moment before entering the house, to shed the daily grind, and keep it outside the threshold.

FRUIT-BEARING TREES, GRAVEL COVER, STONE STEPS AND A WOODEN PATH. YOU DO NOT NEED MORE THAN THAT. A WATER-EFFICIENT GARDEN CAN ALSO BE LUSH, FLOWERING, AND SUBLIME.

A natural pool in the yard is like a

CALMING AND
FLOWING STREAM.

Just as in nature when you encounter a source of living water, the courtyard awakens all the senses. It is possible to feel the breeze, smell the blossoms, listen to the trickle. A natural swimming pool of pure fresh water without any disinfectants or purifiers allows immersion and bathing as in nature; active movement can be experienced against a backdrop of infinite calm and tranquility.

In many cultures, water is a symbol of purity and depth. The presence of a an ornamental fish pond, and even a potted plant of water flowers or small spring from which the water unceasingly flows, are elements that bring welcome tranquility to the garden.

一个独特的、私人的空间

When you use design wisely, to strengthen and emphasize the senses and not just for fashion, you guarantee yourself

A UNIQUE, PERSONAL SPACE.

Allow yourself to roll out the rug, which you have from your very first apartment; to bring the table you loved from Grandma's home; to use only your most comfortable armchair; not to tear down the old roof beams, if there are any; and let the home express who you are. And if you wish, add modern chairs, add you own rhythm and flow—that is your music. The living room is the best place to shake off fashionable trends. Release these concepts when you think of your home, and focus only on what is right for you and your aesthetic.

LIFE IN THE GARDEN INVOLVES CHILDREN, PETS, DOGS, PARENTS, AND GUESTS.

It is a place for a genuine smile, movement, and joy.

THE GARDEN AS A FAMILY PLAYGROUND;
A MAGICAL PLACE FOR FAMILY MEMBERS AND PETS.

Hybrid architecture combines traditional and contemporary motifs

THAT IS HOW THEY WISHED

their home would be.

THAT IS WHAT SHE CREATED FOR THEM.

The house on the mountainside, overlooking the landscape of forests and the sea, was planned in the same spirit as the ancient villages surrounding the house, which were built with inner courtyards. Here, the inner courtyard leads to the yoga room and the front entrance. Connections are built of local stone, concrete, glass, water, and light.

阳台上的冒险

VENTURE OUT ONTO THE VERANDA

TO SEE THE FAR-OFF VIEWS WHERE

in the distance, the sea and sky converge. Ancient terraces intertwine with newer ones. The hammock is a lovely spot to rest and sway. You can also stretch out on the sofas, raise your legs, allow a smile to effortlessly appear on your face. Let the wind blow your hair, let go for a moment.

THE
LIVING ROOM

is the place where nothing

NEEDS TO BE DONE, JUST BE.

起居室

There are few opportunities in modern life to escape to a quiet place where you do not have to do anything. Surround yourself with natural materials and cloth, far from the effects of electrical devices, mobile phones, and the slew of screens that fill our lives. Natural light fills the room from the window over the door to the hallway, providing natural lighting throughout the day, saving energy.

The secret charm of natural ingredients

LIES IN
THEIR GRAIN.
THEY LIVE, CHANGE, RENEW, SEASON.

The materials and items that you choose to bring into the space are what will define it. There is something mysterious and uncharted when you are drawn to certain materials—texture, touch, appearance, smell. The links between them and the memories latent in them are what will set the atmosphere of the room.

There is no need to buy new items to give new life to a room. Start with what you already have;

THERE IS NO RIGHT OR WRONG,
EVERYTHING IS PERMISSIBLE.

We adults, just like children, are drawn to color. Its effects charm us, like a child's painting or a rainbow, a smile appears. Strong colors can bring out joy, gaiety, and good cheer, depending on the context and the boldness.

The use of colorful and surprising textiles will let you demonstrate your taste, testing, and daring. The items of furniture in this living room are based on old pieces, which have been re-upholstered d with great lightheartedness and humor. After all, if you get tired of it, you can always renew and change. High-quality crystal completes the stylistic and nostalgic picture.

用内在意识设计你的家
DESIGN YOUR HOME WITH
INNER AWARENESS

of the best materials for you; their colors, touch, and the sensations they impart.

Imagine the place where you will want to spend your life. Use design as a tool to create this place. Choose materials that speak to you and calm and comfort you. Touch the blankets, the pillows, the carpets; caress them and sense them through touch. Choose the lamp the way you would choose a new pair of jeans or bag; let yourself fall in love with how the light banishes the darkness.

Get to know the soft materials. With your eyes, follow the bird building its nest: watch how it collects twig by twig, stick by stick, and leaf by leaf to create a soft and protected place for its chicks, isolated from the world. What a fantastic understanding of the meaning of a home! These handmade felt items use ancient traditions. It is possible to feel the energy they emit, like a shell collected at the beach. You hear the sea, you feel the creation.

BRING NATURE HOME,

and natural sunlight gently washes the interiors, softens, and in streams a pleasant breeze.

THE NATURAL APPEARANCE IS HONEST AND HEARTFELT.

CYPRESS TREE TRUNKS

piled up on the roadside were the inspiration for this house, which was built by its owners on weekends and holidays

WITH UNPARALLELED INDUSTRIOUSNESS HANDIWORK.

Do you have a vision? Do not take into account what others will tell you. Arrange the home with assurance, based on the feelings of your heart. This integrates your personality into the design of your home. Do you like wood? This is the place to realize your dreams. It can be as in childhood, when you allowed yourself to build a treehouse with the spontaneity, excitement, and passion possessed by children and those who are children in spirit.

How to pick art? Stick to just one rule:

IF IT MOVES YOU,
THIS ARTWORK IS GOOD FOR YOU.

This eclectic living room is in a restored Templar house, brimming with collected design items and intertwined with paintings by various artists from different periods. A renowned artist or an anonymous artist—here too, the choice is always yours. If you are not an art dealer, the best choice of art is the one that will thrill you, make you pause and savor it, time after time. A unique piece of art or design item is like a quality read on the nightstand, with the potential of expanding the human psyche.

A HOME WITH MUSIC

will never be boring.

IT WILL ALWAYS CONTAIN JOY, SENSUALITY, RHYTHM, AND ENERGY.

Music is a universal language, the taste of life. It revives and stimulates, and when you bring it into your home for festive meals and Friday nights, you give a wonderful gift to the entire family. Not everyone has room for a piano or a drum set. Not everyone has to play an instrument, but every home has space for sounds; sing, listen, play, and dance—together or individually, but especially together. A simple radio in the kitchen lifts the spirits and loosens up the atmosphere. And, yes, provide your children with a chance to put on a show—it's what they like best. Join in, because you know it'll be fun.

Children need a place, and not just in a "children's room."

CHILDREN NEED ROOM TO RUN AROUND,

to be together, to live. They are part of life and part of the home. In fact, they are the

HEART OF
THE HOME.

It is important to create for children their own spaces for activities, not necessarily in a designated area or children's room. Moving around is part of their needs, and a home should have a predetermined informal area for the children to run around. It's good for the children, and it's good for the children in us.

A CARPET IS A PLACE.
IT DEMARCATES TERRITORY.

You can lie on it, rest, loiter, or jump about.
Who doesn't need a real rest? Take care to get an especially pampering carpet,
thick and dense, on which you can lie, read a newspaper, or fall asleep.

SENSES,
DO WE UTILIZE THEM IN OUR HOME DESIGN?

Our senses act upon us, influencing our very being:

THE FRAGRANCE OF A CANDLE,

the touch of bare feet on the rug,
the sound of breathing from the one who is lying beside us.

DO NOT FORGET TO UTILIZE YOUR SENSES.

THE LIVING ROOM AS A TRAVEL DIARY.

Living in a distant land has revealed the possibility of acquiring and discovering rare, used items online. Collecting has resulted in long trips, fascinating meetings with complete strangers, and the repairing, building, renovating, and painting of old furniture that comes to hand.

EACH PIECE OF FURNITURE HAS ITS MEMORIES;

EACH PIECE HAS ITS OWN UNIQUE MARKER.

THE AUTOBIOGRAPHICAL LIVING ROOM

The living room is a kind of family album.

IT REMINDS US OF STORIES, TRIPS, EXPERIENCES, AND STUDIES.

Objects collected gradually with love over many years are charged with energy and invite you to simply be. A "natural table" displays a collection of objects from travels across the country and the globe. It is a kind of diary of happy moments, pampering sofas, soft colors, and serene and inviting textiles.

THE DINING TABLE

is one of the most important places in the home.
It is where most of the family meetings are held;
it is where the family is born.

THE DINING TABLE IN OUR CHILDHOOD HOME IS BURNED IN OUR MEMORIES.

We carry it with us always—the sights, tastes, and aromas of our first family.
Whether the meeting was successful or not, the knowledge that sitting down
to eat together is a human urge that transcends cultures and is inherent in humanity,

IT MERELY STRENGTHENS THE DRAWING POWER THAT THE DINING TABLE HAS ON US.

A MEAL IS NOT JUST FOOD.

It is a human ceremonial need to convene and gather around the table,
and the threads that interweave while eating together
all relate the family story of flavors, experiences,

AND BEING A SINGLE ENTITY.

一餐不仅包括食物

"MARKT KITCHEN"

where you can walk around, reach out, and take the washed apples and the Laid-out vegetables, ceramic and wood dishes, and the baskets filled with everything that is good.

The idea of a food table placed in the heart of the kitchen came from Berlin in the 1920s. A collection of old chests of drawers and furniture from villages across England and the Czech Republic comprise the kitchen, creating the atmosphere of a food market where everything is revealed, available, fresh, healthy, displayed, tempting, asking to be taken, tasted, and eaten. All you have to do is to pick up jars of olives or jam, pull out the cheese basket from the refrigerator and the bread from the box, add garden vegetables, and lay everything on a large round wood platter. Time for pleasure.

EVERY HOME NEEDS MAGIC

Ask yourself, what is the mood you want to create? What is right for the interior space is also right for the exterior space, the yard, and veranda. Shape these magical moments in the home, the moments of togetherness and the moments of solitude; intimate, informal, unpretentious, simple places that are available for relaxation. They should have it all: the movement of a hammock, the serenity of an armchair, the magic of a bench, and a place to eat together.

A KITCHEN OF LIGHT, AIR, AND HEALTH

A healthy kitchen is a kitchen flooded by controlled natural light,
and air that flows from the open window carrying the aromas of stews to the outdoors.

光之厨房

A COMPLETE AND COLORFUL GARDEN INVITES YOU TO ENTER

its embrace and set out a table under the fruit tree.

THE HOME IS A SINGLE ORGANIC UNIT;
ENTER, LEAVE, AND FEEL THE CONTINUITY AND FLOW.

The paths in the house run from the kitchen outward to the veranda,
up to the second floor, to the bedrooms with views to the stunning vista.

THE KITCHEN AROMAS FILL THE ENTIRE HOME, LIGHT FLOODS IN WITH EASE, AND NATURE IS ALLOWED TO COME INSIDE FROM EVERY DIRECTION.

THIS VERANDA OVERLOOKING THE VALLEY PROMISES SO MUCH.

A pair of rocking chairs on a porch is always an invitation to go out and breathe in the wide open spaces.
The accessible kitchen, close to the outside dining table, turns every meal into an ode to the view.

NATURAL MATERIALS ARE THE BASIS OF LIFE.
An inviting kitchen summons you back home to eat home cooking,
with the aroma of baking issuing from the oven, and fruits and vegetables fresh from the garden.
ADDING TO THE SERENITY OF THE KITCHEN,
THE FOOD IS HERE AND IS PLENTIFUL.

THE GENEROUS
AND CREATIVE KITCHEN

Fruits and vegetables from the garden are laid out and prepared in a way that invites them to be taken and tasted.

THE AROMA OF BAKING FOCACCIA EMERGES FROM THE OVEN, FLOODING THE ENTIRE HOME, COMFORTING,

despite the burden of history and memories of a 100-year-old Templar house, with all its landmark preservation laws and meticulous reconstructions, and despite the weight of the stones that tell an ancient story—it is a "house of life." The atmosphere in the kitchen is refreshing, thanks to the spaces, which have been adapted for every activity: a place for independent cooking; an area for baking pita, focaccia, and casseroles; a separate area for peeling, cutting, and slicing. Everything invites a joint effort by the entire household.

GO OUTSIDE, INTO THE FRESH AIR

呼吸新鲜空气

The veranda is another part of the home, an outside room. It invites you to go out, continue your routine, greet the morning with the first cup of tea, breathe in the scent of the jasmine, and hear the birds. A conversation begun in the kitchen can continue in the breeze. There is something about a veranda that makes it possible to change the climate, refresh, and breathe.

MOTHER EARTH
ENCOMPASSING, NOURISHING, OVERSEEING, THRIVING

The vegetable garden is an excellent opportunity to develop in your children a sense of responsibility for nature, and offers something in return. They can enjoy following the development of the invigorating tomatoes, and the eggplant that suddenly sprouts; they pick the juicy peppers and cucumbers for the next meal. The backyard has a wide patio and a well-tended vegetable garden, gazebos, orchards, and a vineyard that surrounds the house. There is almost no need for a lawn, which guzzles water and requires constant care, but if you feel the need for a green platform, it is a kind of outside carpet on which you can rest or play.

THE HEALING ENERGY OF THE GARDEN

How powerfully healing it is to stroll down the garden path. To rest on a bench and listen to the sounds of nature, the wind, the rustling of leaves, the flapping wings of the butterflies. The gifts that a bench like this can provide are precious, and should be savored often.

THE GARDEN INVITES YOU
TO LINGER AND DELIGHT IN ITS BEAUTY AND INTRICACY.

An assortment of settings is formed between it and the home.
A dense garden surrounds the house; walking in it is like walking out in nature.
The paths diverge and shift; one rises up, another slumps down. Stone steps or a path of fine gravel beckon, and the plants are particularly lush and wild.

THE GARDEN IS NOT REVEALED IN A SINGLE GLANCE.
IT INVITES US

to venture out into its hidden distances, to hike along the mountainside, in the ever-changing, complex topography that contains different corners and rest areas.

WHAT DRAWS US TO STAY AT THE TABLE AFTER THE MEAL HAS BEEN EATEN?

What compels us to sit in the kitchen even when eating is not the goal? Why do children congregate there to do their homework? The emotional, intimate, enveloping design frees the kitchen from the usual appearance,

TURNING IT INTO A NATURAL CONTINUATION OF THE LIVING ROOM.

是什么让我们在餐后留在桌边?

The cabinets, carpenter tables used as workspaces, open shelves for display and storage, living room–style light fixtures, utensils that have been passed down the generations, and a single artwork together create a kitchen that does not feel like a kitchen. It is a kitchen that blurs every border on the one hand, and creates a comfortable, organized, and pleasant area for dining for the whole family on the other hand. It is important that the kitchen works for you. If brown calms you, use it; if you need brighter colors, don't hesitate. Your kitchen should be comfortable; everything begins here.

THE DINING ROOM TABLE MAY BE THE HOME'S CENTRAL ELEMENT.

Entire lives are lived around the dining room table; it attracts a flood of events, people, and situations. It shapes our childhood memories, and will set the aroma and memories of our children; setting the table, feeding the diners, the good and not so good conversations. Eating together means to take trouble, prepare, cook, set the table, serve, mix, pour, pass around, clear, wash up, clean up, and arrange—the fixed rituals of a family and friends.

PUTTING TOGETHER A PRECISE KITCHEN IS A LOT LIKE COOKING.
IT'S ALL ABOUT PROPORTION.

CONCOCTING COLORS AND TEXTURES

and finding the right connections that get the job done—it is just like quality cooking. Use surprising ingredients that will cause guests to pause and wonder what is in this dish, in this kitchen, that is so riveting?

EVERY DINING TABLE IS A SOURCE OF EXPERIENCE,

ATMOSPHERE, STORIES, FLAVORS, AND MEMORIES.

Seek the suitable corners, either hidden or open, in the home or yard where it is possible to place another dining table for either an intimate meal or a larger group. Seek shade, inhale the breeze, and use nature's energy wisely to create places with disparate characters for different events and diverse experiences. This is the time to think about the desired seating: around a round table, on a wooden bench, or on upholstered chairs?

The atmosphere around the dining table is supported be the space in which it is found: an artistic creation painted on plywood, a collection of cabinets and renovated tables, a tiled floor taken from an old house, and delicate, colorful light fixtures.

The design affects the

MOOD, ENERGY AND VITALITY.

HOUSEHOLD MEMBERS PRACTICE CHANGES

and movements, serenity and a relaxed flow. A pleasant setting

FOR A SENSE OF SIMPLICITY.

The kitchen is a wonderful place to
SPEND TIME TOGETHER.
It is the center of the home, a place to prepare a meal, set the table, eat together on a daily basis; to foster a conversation and share with one another, even in trying times. To feel like a family. And in families everyone has responsibilities—both in preparing the meal and in cleaning up afterwards. Early involvement prevents unnecessary frustrations

WISH FOR A MOMENT OF VACATION IN THE MIDDLE OF THE DAY?

A dining table in the courtyard is a wonderful place to create a peaceful vacation-like ambiance. When deciding on the position of the table, it is important to ensure that there is a proper air flow around it. Such a table is exactly the place to relax with a cup of afternoon tea, or a glass of wine in the evening. Your own place to relax and

SLOW DOWN.

A CALMING GARDEN, INVITES YOU TO ENTER IT.

Choose a corner, and devote yourself to

THE SILENCE OUTSIDE.

Arranging the tables differently offers a different experience in each corner:

sit alone, enjoy a meal with many diners or nestle in for a romantic breakfast for two. Abundant flora, rich in color inspired by landscape paintings,

GIVES EACH AREA A PLEASANT ATMOSPHERE.

OBJECTS PASSED FROM HAND TO HAND BETWEEN FAMILY MEMBERS

were grouped and joined together alongside objects that had been upgraded, renewed and repurposed into something else. The colorful carpet serves as the center of the dining room, attracting most of the attention. The family members gather together, enjoying nostalgic stories, memories sparked by the old objects, alongside

HAND-CRAFTED ONES.

TAKE OUT GRANDMA'S SERVICE, MOM'S TABLECLOTH AND YOUR HUMOR, GOOD SPIRITS AND FREE-FORM THINKING.

Extend your creativity to

THE DINING TABLE.

The sense of protection and security offered to us by the house can also exist in the inner patio courtyard, which is protected on all sides by the walls of the house. Such a patio provides protection against high western winds. In general, a terrace is adaptable, allowing us to alter the atmosphere often, and holds the promise of surprise.

Cooking for family and friends is an opportunity to convey a **DAILY REMINDER OF SHARING AND LOVE.**

The temperature in the kitchen affects the experience. Proper planning—allowing for heat control, natural daylight and artificial lighting, natural airflow, and air ventilation—can dramatically improve the feeling in the kitchen, and turn an hour of toil into an hour of pure pleasure.

THE SOUL NEEDS SPACE,
THE BODY REQUIRES PHYSICAL MOVEMENT,
灵魂需要空间

the eye longs for the horizon, the heart yearns for family, for friends, for the feeling of togetherness. With the addition of a pleasant breeze, the touch of freshly cut lawn and the smells of blossom,

IS ANYTHING ELSE NEEDED?

Light reflects on the bright wall colors, on the handmade ceramics, on the wooden platter, on the iron chair and on the stone floor.

LIGHT ALSO REFLECTS ON THE FRESH BOUQUET OF FLOWERS PICKED FROM THE HOME GARDEN

and, figuratively, on the mood of those around.

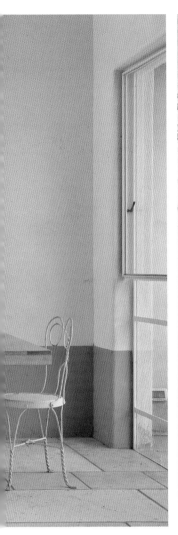

HOW BENEFICIAL IS YOUR LIGHT?

WHAT LIGHT DOSE IS RIGHT FOR YOU?

Is it possible to control the natural light that flows into your house? Light is one of the most significant elements in setting the tone around the table. Natural light, candle light, artificial light—all affect the space.

什么样的光线最适合你?

The kitchen cabinets are softly awash with natural
LIGHT PERMEATING THROUGH THE WINDOW.

LIGHT AND AIR

are key elements when creating work conditions in the kitchen:
with precise and economical design, the cooking experience can be notably enhanced.

SOMETIMES COURAGE IS REQUIRED TO DARE

and place things differently, to muster surprising combinations,

AND USE NEW MATERIALS.

有时勇气要求敢做

The kitchen is a great place to showcase our creativity while cooking, organizing, decorating, and maintaining proper habits of work, cleanliness, and order. Like in cooking, design sometimes it takes courage to arrange things differently, to surprise with unfamiliar elements, and to shake off conventional and preconceived notions. Such design (and cooking) will cause the observer to stop for a moment, explore the details, and delight.

TO LIVE IN VIEW OF THE MOUNTAIN, WITH SUCH A VISTA, CREATES A RELATIONSHIP.

Determine for yourself how you wish to take in the landscape:

how much exposure you are comfortable with opposite the infinite views?
Experiencing the ridge and open space from

WITHIN THE HOUSE

creates a deep affinity with nature. The landscape—near and far—fits into the topography.
The home's vegetable garden provides plenty of organic vegetables to the kitchen on a daily basis.

Just as in the biblical tent, where the family dwelled in one common space, where they cooked, slept, lived and kept warm in the light of the burning fire in the center, so it is in this house, made of

CONTEMPORARY MATERIALS.

The shared open space provides security for children, especially toddlers and young ones who seek the company of adults. Grown-ups, too, feel more confident when they are in control and can monitor what is happening in the area, maintaining eye contact with the children even while in the kitchen. The wooden house was built by the owners from a deep-seated conviction for minimalism that advocates for minimal use of objects and accessories.

THE NATURAL LANDSCAPES

present beyond the door contrast with the softness of

THE HOME INTERIORS,

creating a dialogue between the interior and exterior. One is protected and one is exposed, but they stand together, connected in the world, and converging.

THE POWER OF STYLING, OF HOME DRESSING, IS POTENT.

It is choosing the dark tone of the armchair, the one that household members particularly love, that compels them to curl up, feel wrapped and protected.

THE GOOD MOOD KITCHEN

RED KITCHEN CABINETS, A GREEN BENCH,

the painted tile floor, and a renovated armchair make for surprising combinations of colors, furniture pieces, accessories, and style.

幽默、色彩与欢乐

HUMOR, COLOR, AND JOY
CAN NEVER HURT.

THE COLORS

evoke the mood in the kitchen in a variety of displays: the small electrical appliances, the textile of the map, the dinnerware,

AND THE MENU ITSELF.

If you feel happy (and rightfully so) with a dining table that serves both everyday food as well as festive holiday meals, be sure it's generous and fits the various activities taking place in your home. Your tastes, colors, and cheer can additionally be added into the kitchen via colorful dishes and accessories.

The wooden floors both inside and outside the house create continuity and flow between the outdoor and the interior, between the near and distant landscapes. The wooden veranda overlooking the expanses is made of natural organic materials, suspended above the ground, simulating a sailboat deck. The use of natural wood represents growth, regeneration, and integration with the natural surroundings. One somersault and presto—they're out of the house. Spontaneous, daily, joyful transitions. Plan the landscape and the space, imagine life's many varied moments that will take place in it. Recognize that reality goes above and beyond any imagination. Sit on the veranda, looking out over the vista of fields. During the summer, field crops ripen; harvest begins in preparation for collecting the crops from the fields.

STRETCH OUT ON THE COUCH
AND SAIL INTO AN INFINITE VIEW,

right here from within the house. A landscape that continuously changes with the hours of the day and the change of seasons resembles works of art that, one by one, transition on the wall.

EACH OF US HAS OUR FANTASY OF THE PERFECT SPACE.

WE ALL NEED QUIET TIME.

我们都需要安静的时光

Designated transition spaces that separate the public areas from the private quarters are wonderful spots that can provide you with some much needed quiet time. Comfortable, inviting armchairs with plenty of cushions, located in surprising areas that guests do not always have access to, will undoubtedly inspire a release. A soft, pampering sofa in the middle of nowhere is a great place to curl up, read, listen to music, or simply sit silently.

A passage, a pause, an interlude.

AN INVITATION TO REST

just before entering the bedroom.

A SPACE SEPARATING THE BEDROOM FROM THE REST OF THE WORLD, PROVIDING IT WITH COMPLETE PRIVACY.

The master bedroom is the parents' exclusive retreat, their own intimate sanctuary. They are welcome to close the door, request privacy without upsetting anyone. The private bedroom offers complete indulgence. The view towards the eucalyptus grove, the delightful bath, the generously wide bed, and the narrow veranda allow space to pause and breath. The position of the room in relation to the direction of light and wind is immeasurably significant, influencing the climate and mood in the room. The bedroom offers a promise of comfort and pleasure, alone and together, with or without the bathrobe.

THE RENOVATED TEMPLAR BEDROOM PRESERVES THE SPIRIT OF THE PLACE,

respecting its past while creating a luxurious, pampering living environment that is connected to the present.

On the second floor, at the end of the hallway, is the intimate bedroom with its 100-year-old wall. The wall was exposed and treated for conservation purposes. Old wooden doors have been restored and installed, and the few furniture items were carefully selected, as to not create a visual load. The use of natural materials such as wood, stone, and plaster provides the room with its rustic atmosphere.

WAKE UP INTO THIS MOMENT,

THE WINDOW OPEN,

the air flowing in softly. The curtain flaps gently and the sun draws delicate lines on the white linens.

IT MAKES YOU WANT TO CURL UP,

WARM IN THE SILENCE.

JUST LIKE A GREAT MEAL,

THE BEDROOM TOO, CAN CONTAIN GOURMET ELEMENTS.

Fine bedding,

THE PERFECT LIGHTING,

THE DELICATE CURTAIN, A HAND-PAINTED BED FRAME,

A ROMANTIC BENCH

a wondrous frame you can step into,

AND MAKE LOVE IN.

A bathing experience in

SHADES OF LONGING.
HOW WONDERFUL ARE THE POSSIBILITIES INHERENT IN COLOR

—with the shades, textures, and impact it delivers. Knowing that it can be easily replaced at any time with the collective effort of family or friends; knowing that it empowers, delights, and liberates emotions. Water, color, and moonlight. How wonderful. The bedroom as a sanctuary detached from daily hassles and distractions. Knowing that nature is a monumental source of energy, the more it is present inside—within ourselves and our home—the stronger we grow, brimming with passion and love. The bedroom as a place for caressing, touching, and fulfilling. A place for excitement, admiration, and harmony. It offers a different level of giving, and a generous place for love.

THE BEDROOM FEELS LIKE A TREE HOUSE,

A WORLD HIDDEN

from plain view: concealed, personal, and intimate.

The intuitive, functional design is in tune with the emotional state of the tenants. The poetry books lying next to the bed; the soap she likes, his reading lamp: these elements that convey who they are, what they love. How authentic and cozy this bedroom is, like a wooden house made of durable, natural material: decades-old oak floor. Natural lighting, a sea breeze, renovated doors, and old furniture complete the gorgeous setting.

CREATE A WONDROUS SETTING

FOR YOUR LOVE AND RELATIONSHIP.

SET UP RELAXING LIGHTING, A PLEASANT SPACE, LUXURIOUS UNDERGARMENTS, NATURAL LIGHT,

and air circulation. Use fine products, quality linen and basic furniture pieces that will keep the clean and pristine appearance of the room.

BATHING OUTSIDE THE ROOM,
IN THE WARM AIR,

is a true desert experience. The sky and stars are visible through the straw arbor, as is the sun. An outdoor experience, yet protected from sight. Privacy with a wonderful sense of freedom.

THE PURITY OF SPACE,

the shaking off of congestion, making do with little, leaving room for thought, containment, not overloading objects or furniture: they leave space for new things to enter, particularly clarity of thought,

BUT ONLY TO A DEGREE.

The bedrooms are located on the top floor of the house,

SEEKING SERENITY AND FREEDOM

from crowds and clutter. They reveal a minimum of objects,

FURNITURE, AND DETAILS.

The joy of having a little is like liberation from dependency. Let the structure be present in clear, minimalistic lines.

THE WALLS, LIGHT, AND THE SPACES

are the actors. Wooden doors that stretch to the ceiling leave room for the light to enter, allowing it to flow freely from one area to the next.

The room with a door you can always close,

FREE OF ALL OBLIGATIONS

FOR A MOMENT.

Read, write, disengage and enjoy guilt-free privacy. When you need a quiet corner to escape to, this is the place. Some prefer to place the writing corner, laptop, TV or audio system in the bedroom; others prefer to leave all digital devices outside the bedroom to maintain maximum tranquility. This room is designed to serve your preferences; do as you please.

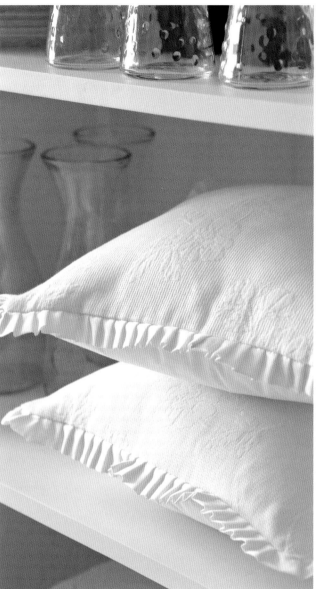

PILLOWS, TOWELS, TABLECLOTHS, NAPKINS, CURTAINS, AND TEXTILES

soften and fill the room with pleasantness.

纺织品

Soft, white, high-quality, hand-sewn textiles, which with just a light touch make it possible to feel the quality of the texture: crude or delicate stitches, accentuated texture, thick towels packed with satin thread. Touch the material, feel its soft power, caress every possible crevice, snuggle in, be pampered, and absorb.

 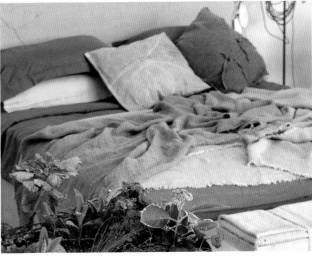
 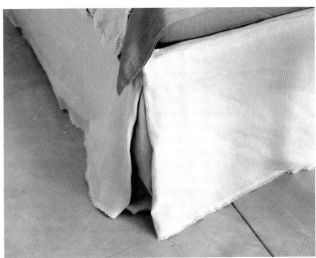

THE MAGIC OF SPACE

can lead to wonderful combinations of

MATERIALS IN VARYING COMPOSITIONS.

Connections of contrasting materials together create a subtle harmony. Define the space and let it be. A glass beam adjoins a concrete one and flows onwards, to a plaster beam and a glass display case. There, it converges with the local masonry wall, ultimately intersecting with the exposed concrete wall. Fascinating juxtapositions of materials, light crevices, and small furnishings create an inviting, enabling space.

THE PASSION

is present in movement, in the layers of textiles;
the hearty light floods the room, inviting you to leap
into the bed, to curl up, go wild, make love.

LAYERS UPON LAYERS OF GLORIOUS,

indulgent textiles in varying hues. From the rug and the bed skirt,
to the myriad cushions. Each of the textile elements is different
yet similar, blending together, just like us.

IN YOUR OWN PRIVATE KINGDOM

you are allowed to go wild with colors, and choose textures and materials that only you adore. Does it fit the rest of the family members? Do they enjoy the result? Don't think twice. Your private bedroom and bathroom are the places to fulfill all your fantasies: the humor, the color, the joy,

AND PERHAPS A SURPRISE ELEMENT.

家庭思维

At the end of a long journey or an arduous day,
all you dream of is coming home,

THROWING YOURSELF ON THE BED AND CLOSING YOUR EYES.

After a shower and some freshening up, it is possible to shed
the intensity of the day and step into another, more relaxing mindset:

A HOME MINDSET.

The bedroom serves as that transmission station
where one takes and sheds form, ridding oneself of the day,

OPENING A DOOR TO THE NOW.

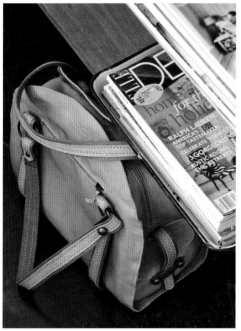

TO RECLINE IN FRONT OF THE VIEW, savor the heat of the fireplace, turn the bed in the bedroom to THE HEART OF THE MATTER.

The more we contemplate our bed's positioning, the more pleasurable the experience we can have .The bed's position on an elevated platform enables one's head to rest on the pillow while at the same time observing the views, the living room level, and the garden. Nothing obstructs the spectacular sunset and sunrise that can be seen from here, only the pergola was constructed specifically to block the sun's rays from entering and interrupting a sleepy waking, a vacation-like awakening.

Is there such a thing as a bed that is too indulgent?
We need a good, deep, and

BLESSED SLEEP.

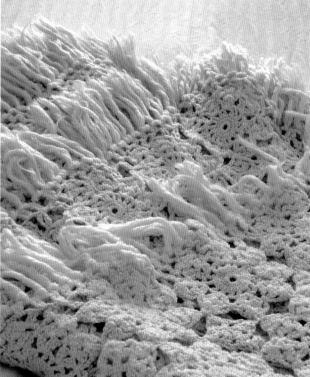

We need a place to fall asleep with a good book or after an intimate conversation, and to wake up in. We need this time, sleep time, in order to continue our daily work. Well-cared-for and padded surroundings are all that we need. And light; natural light is as essential as air, food, and water. Our lives are organized by the hours of the day and night, by which our body clocks tick.

THE ENERGY
FELT IN THOSE SPACES CRAFTED BY HAND IS INEXPLICABLE.

It's magic. An old pharmaceutical chest made of drawers was repainted. The cushions are made of handmade textiles, and an armchair that was purchased in the flea market was renewed.

Leave space for pets.

THE FISH IN THE AQUARIUM,

the cat lying on the rug, the beloved dog who sometimes stretches happily on the bed. Caring for pets is an extremely important task for our children;

IT TEACHES THEM RESPONSIBILITY, DEVOTION, AND LOVE.

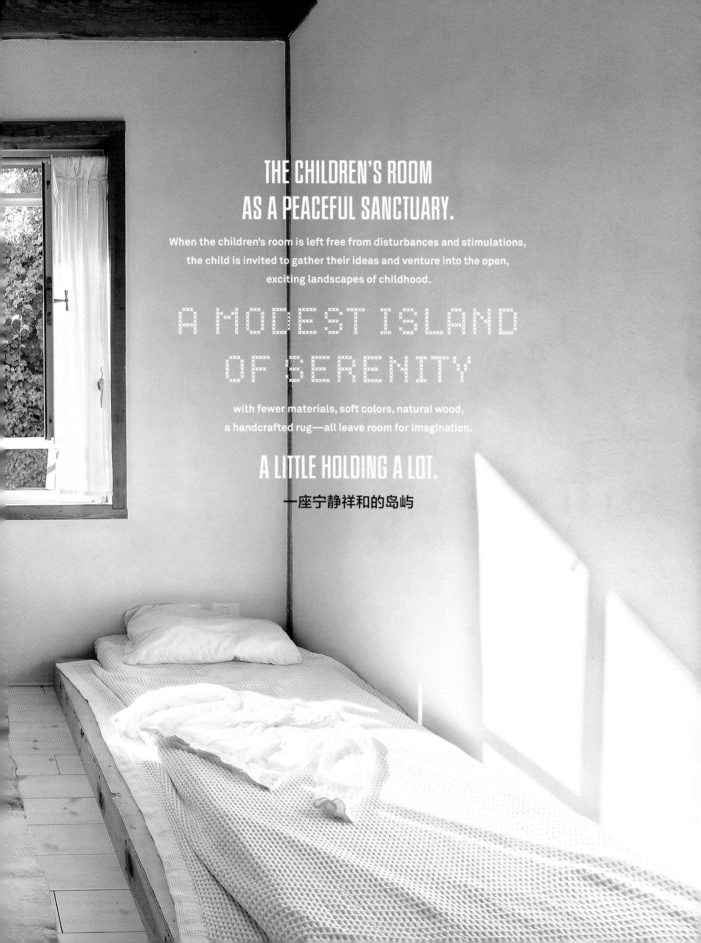

TOGETHER OR SEPARATELY?

The most important question when organizing children's rooms is
how they prefer to live and what you, the parents, think.
Should the children share a room or have, if possible,

SEPARATE SPACES DESIGNATED FOR THEM ALONE.

Dedicate different areas in the house to them,
areas that are outside their room,

AREAS WHERE THEY ARE ABLE TO EXIST TOGETHER,

freely and comfortably. If you define such areas,
it will be easier for everyone to live side by side,

TO CONNECT,

AND TO KNOW HOW TO DISENGAGE WHEN NEEDED.

Tactile sensation greatly affects children and

INFLUENCES THEIR SENSES.

The materials found in their room may cause mood fluctuations. Subconsciously, when they say, "I don't like it", they know what they are talking about. Crafts made from natural materials can convey a

STRONG ENERGETIC SENSATION,

simply by knowing they were handcrafted lovingly. A handwoven rug carries a simple memory of a gentle touch.

Color is unquestionably one of the elements that most influences

THE SENSE OF A ROOM.

Color influences everyone differently. It is a good idea to check with your children which colors they prefer, which colors make them feel calm.

DEEP BLUE

CAN CALM AND OFFER A SENSE OF SECURITY.

一个房间的感觉

The collection of simple materials from which the furniture is made is

THE SECRET THAT

CHARMS THE ROOM.

wood boxes, simple plates, and boards are the basis for preparing the furniture, combined with an old kitchen cabinet for storing the toy collection.

NATURAL LIGHT FLOODING THE ROOM AFFECTS OUR DISPOSITION,

our mood, who we are. Light reaches us in various ways and can create a thrill, a sense of relaxation or focus.

A collection of tin cans, furniture that was updated and painted white, an inspiration board, a soft rug, and the right tone of color on the wall create

HARMONY IN THE ROOM

and unfold the child's story:

HER SECRETS, HER COLLECTIONS, HER AFFECTIONS.

THE CHILDREN'S ROOM

in the conserved and updated house utilizes its original structure and materials. The window, the wooden ceiling and the floor tiles were dismantled from another room in the house and brought here.

The children's room is like a

CANVAS FOR LIFE.

It allows them to delve inside and continue to grow inside the new creation,

TO DEVELOP AND EXPRESS THEMSELVES.

DELIGHTFUL OBJECTS, SURPRISING COMBINATIONS, A STATE OF MIND

A cheerful chest of drawers painted bright turquoise; an embroidered bag purchased in a distant market hangs on a shutter that was picked up from a pile of disposed renovation items in a big city. A treated wall is painted gray at the bottom, and on the top shows wallpaper with flowers. A painting of surprising proportions in pastel hues, albeit overbearing, also evokes a reaction. You cannot remain indifferent—and that is great.

HOW GRACEFUL THE SIMPLE THINGS ARE,

how much emotion they express.

Sometimes you just want to touch them, to feel them, to inquire—who crafted the ottoman? Where is the rocking horse from? The closet? Who knitted the rug? There are so many details that tell stories of journeys, connections, and love.

A renovated desk, an inspiration board made of a window frame, a basket through which old fabrics are woven into knots, an abacus made of recycled corks and arresting hand-knitted ottomans. Here is a wealth of ideas, colors, and cheer, with elements that add grace and individual style to every room. Use them, encourage the creators. Reach out and engage with the elements. It's therapeutic.

HANDCRAFTED OBJECTS MADE LOVINGLY

will always contain an unexplained energy, instantly linking

EMOTION AND FUNCTION.

THE GIRL'S ROOM IS ORGANIZED AS A PRIVATE SPACE,

where all little moments special to girls only are held: collections arranged on shelves, fringed and tattered pillows, and countless secrets. Stop for a moment and organize the room: what exactly does she want to place in it, what colors would she like,

AND WHERE SHOULD HER REST AND SLEEP?

A BABY ARRIVES HOME. THERE IS NO GREATER EXCITEMENT.

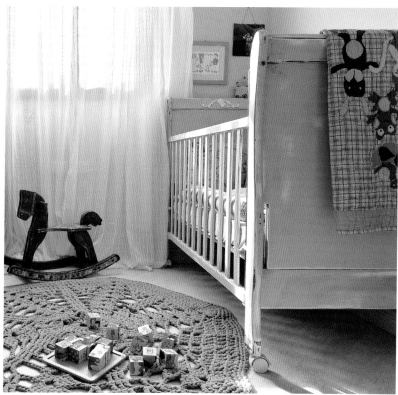

Every family member gathers to receive the newborn baby onto their lap,

MAKING ROOM FOR THE BABY IN THE HOUSE AS WELL AS THEIR HEARTS.

Imagine the baby, swaddled in a small, soft package,
carefully placed on the bed that was lovingly prepared in advance.

Fabrics enable us to play with various color possibilities: to inject joy and exultation into the children's room and, when we feel the need, to remove it and move on, to a new era, another age. When a cork floor meets a mountain of cushions and seats, it instantly

INVITES US TO SIT DOWN.

The desire to create a room for our children that they will adore is connected to the desire we have to give them the wonderful world that was, or was not, a part of our childhood. We collect, find, and arrange

ALL OUR DREAMS

in order to create a place that will fill them with joy and confidence, that will accompany them through all stages of their growth.

CHILDREN LOVE WATER,
BATHS FULL OF FOAM,
THE SEASHORE.

THEY ENJOY GETTING WET AND PLAYING AROUND,

and we cannot be more delighted than when beholding a bathed,

CLEAN, AND FRAGRANT CHILD.

There is nothing like a dip in the water to start off or end the day. Children love water,
they love the blue color and have a natural thirst for life.

The transition from the bathroom to their room is always accompanied by

CHEERS OF JOY.

Sometimes it takes a great deal of effort getting them into the bath,
but also no little effort getting them out of it.

居家办公
WORKING FROM HOME

is not only a matter of convenience,
it is also an environmental and social agenda.

When we work from home, we save time, fuel, and electricity, and avoid missing out on family moments, small and large. Sometimes it is a studio in the bright, well-lit basement, other times it is a space in the yard or a repurposed chicken coop that becomes an artist's workshop.

THE LIGHT ROOM

A LATE ADDITION TO THE CONSERVED HOUSE

Out of respect for the preserved house, and so as not to damage it, the idea of a light room was born, as an addition to the original structure. It's a narrow, built-in unit with huge windows that lets the light flow in and where the study is also located. You can sit down across from the view and enjoy the gardenscape surrounding the house, as well as the side garden, which is exposed through a small window set between the library shelves, just as if it were a framed painting.

THE POTENTIAL
FOR WORK CORNERS IN THE HOUSE IS
hidden and revealed in surprising locations.

Sometimes these are areas that we planned, defined, and designated for work, other times we did not plan them at all. In the age of the laptop, wherever we sit down comfortably—there is our work space. Life dictates conduct. If we are content sitting amid family members while working, the dining table will do; if we need peace and quiet, and so does the rest of the household, we will seek more remote, hidden places further away.

Kitchen-studio-home

A FAMILY CONTINUUM
UNINTERRUPTED

Advanced technology connects us to the world from just about anywhere; even coffee shops are offices when needed. In the morning, the house is emptied of children and the active studio awakens for another day. A lowered wall and dedicated storage solutions are all that is required to combine creation with life.

CAN YOUR HOME
ALSO BE YOUR WORKPLACE?

The desire to work from home is growing as our technology becomes more sophisticated, from the computer terminal to the smartphone and tablet, which allow us to connect to the world from anywhere. Cafes are places to hold business meetings and office work is done from home. Even if we cannot clear a separate room for this purpose, it is still possible to have an office or studio in an apartment's common area. All that is needed is the right organization of family activity (and the knowledge that the home is empty of children until the afternoon), and planning to offer multiple storage solutions. A work corner behind a half-wall offers privacy and concealment from the living room.

The household tends to call this The Room.

Use of the capital "R" merely accentuates the importance of this space as The Place. It makes it easier to host, create, repair, and store. The Room is a dynamic space. It is where we keep bags and boxes, where the work spaces are, where we store and do the laundry. It is where the house's chest of drawers is located, once used by the Tel Aviv University library and now used as the main organizer of the home with a place for everything: the shears, nails, screws, candles, and what-nots. There are so many things in every home that need multiple practical and functional storage solutions.

WHEN THE DESIGN IS AUTHENTIC,

derived from function and need,

IT WILL ALWAYS BE RIGHT AND SUCCESSFUL.

Need is the mother of invention, and the invention is a graceful and well-designed object. How much space should be devoted for shoes? How will we hang the laundry inside? Where will we store all the tools? The books? The rubber bands? How will we organize space for everything, especially the little things? This is part of sustainable design, which adheres to an ecological agenda, which does not seek to replace the grater merely because the trend has changed. Everything is thought out and based on a real need.

A renovated and restored stone storeroom where every stone is collected and reused with the sole purpose of not throwing out the history, but instead renewing the ruin and connecting it to the house. It shares the same philosophy as re-upholstering the armchair, renewing the light fixture, and changing colors—to update and continue life.

LIKE A WALK IN A FIELD
仿若漫步田野

Bathed in a richness of flowers and plants, a small garden also needs advanced planning to fully utilize its space. The planning process is always an act of learning and listening to the needs of the family. Planning of a garden on the grounds must consider the wishes and dreams of the people who will use it, with attention paid to colors, materials, styles, and tastes. For me, the garden connects the different actions that will be carried out there.

A whole palette of colors is gently present in the living spaces, the walls of the building, the details of the furniture, the yard, and the garden. The cushions, the sofa upholstery, the shades of the front doors, of the curtains and the rafters, all together create a

WONDERFUL HARMONY,

as if the place is stirring with joyful music.

ONE OF THE PLANNING GOALS WAS TO ACHIEVE MAXIMUM CLIMATIC COMFORT.

To cope with the low afternoon sun, a portable system of wooden lathes was designed, providing shading as needed. The interior of the house is particularly thick with concealed shutter boxes designed to improve thermal insulation. If you are willing to leave city pleasures for a land plot with infinite views, then connection to the air, light, and nature flows in your veins. And when nature is present with all its might in all corners of the home, it is important to provide climatic solutions that will merge and balance between the interior and exterior. This house uses green building principles with an emphasis on an innovative, geothermal heating system that uses the soil as a source of heat exchange.

THE SEASONS CHANGE,
LEAVES ARE FALLING SLOWLY, HINTING AT THE COMING OF FALL.

The mood on the first wind-blown days,

THE SKY'S COLORS ARE CHANGING, REFLECTING IN THE POOL.

The mood—a lyrical whistle—is heard down the stairs;
someone recalls a romantic tune and continues out, hopping out to the yard.

WALKING THROUGH THE GARDEN PATHS
BRINGS TO MIND AN ETERNALLY OLD, FAMILIAR PLACE,

reminiscent of ancient times—of an age-old Galilean landscape.

It seems that the delicate, calculated design conceals layers upon layers in order to evoke the sense of natural wilderness. Planning the garden's various components begins as early as the development phase; the construction of the rockeries, laying the ground coverings, and planning the advanced irrigation systems—all are done with minimal future maintenance in mind. Saving resources, water, and energy is the essence of environmental planning.

SOMETIMES ALL YOU NEED
IS TO VENTURE OUT AT
THE RIGHT TIME TO

CAPTURE
THE MOMENT.

绿色假期

A GREEN VACATION

When the body begs for a break, the soul asks for renewal,
and the mind seeks clarity, it is time to go on vacation; discover a

UNIQUE EXPERIENCE

that will restore all that has been lost under the intense pressures of the everyday.

CONNECTING THE NATURAL,

THE RELAXED,

and the primal elements enables the body and soul to find once more
its own rhythm and be filled with fresh, renewed energy.

Amid the abundance of health, beauty and cosmetic treatments, gourmet meals, cultural events, and diverse activities, all bathed in the scents of jasmine, lavender, and lilac, guests are invited to relax, rejuvenate, be moved, leave behind the daily hassles, and reconnect with themselves. The bright rooms overlooking the landscape let the light breeze carry nature's scents inside.

The hotel overlooks

THE MOUNTAIN RANGE
AND THE LAKE,

surrounded by infinite tranquility,
greenery, and bird song.

The hotel's 37 acres feature meticulously designed suites and rooms, an organic farm and farmstead, orchards, herb gardens, flowing water, and everything the body and soul could wish for to enjoy the perfect harmony between humankind and nature.

THE BEDROOM

is open to a wraparound veranda that extends the connection to the

LOCAL FARM, THE LANDSCAPE, THE SCENTS, AND THE AIR.

This site is home of one of the world's most diverse organic farms, cultivating fruit and vegetables, alongside a hutch, a cow shed, a poultry shed, and a dairy farm. Both cows and sheep graze freely in pastures. The restaurants on the premise make use of the organic seasonal output from all parts of the farm. Leftovers are not tossed away but are used to create naturally processed compost. Glass and plastic bottles are sorted and collected. The farmstead output is used to create a wealth of natural products including dried fruits, jams, liqueurs, infusions, and soaps. The air carries a promise; fall is around the corner. The trees will soon shed their leaves in preparation for winter. We, too, are preparing to withdraw, to gather together. In the warm house, we prepare the wood for the fireplace; the soup is brewing on the stove. A sense of longing arises. Emotions overwhelm us. We slowly gaze that quiet, peaceful gaze. Sadness, too, has a place. One can stay in it for a moment or two.

"We created the place

WE HAVE DREAMED OF FOR OURSELVES,

and we are delighted to share our experience with our guests."

Nestled amid the green mountains is this house where you will find a different kind of vacation experience. On the bank of the stream, in the heart of a wild grove of oaks and pistacia trees, the house is built entirely of natural materials, pleasantly painted, and connected to the generous space of the groves. It preserves the harmony between nature and humans, between two people, and within an individual. The house was built by a couple, designers and partners, with a passion for realizing their love of beauty, harmony, and hospitality, and a wish to share it with those who appreciate it. The house and suites are located in a natural garden of Mediterranean forest, along with a system of living pools that influence their surroundings.

The passageways within the house, the guest rooms,
the garden paths, and the verandas are

THE TRANQUILITY ROUTES OF THE PLACE, AND THE SECRET OF ITS CHARM.

The guest rooms are each organized as private accommodation, allowing guests to completely detach and let go. The smooth and fragrant linens envelop an exceptionally wide bed. The special mattresses provide flawless rest and sleep. The towels are soft and large. Bathrobes and an indulgence kit are a splendid treat. A private collection of furniture, tools, and objets d'art, the fruits of global travels, reveal themselves slowly.

THE HOURS BETWEEN
THE DAY AND THE NIGHT,

the moments at dusk when the sun descends
into slumber, its final rays adorn the

sky's expanse.

As it sinks, it radiates onto the pond turning the water into what appears to be a dark-colored lake, it paints the green surfaces with magical shades. At sunset, we look out of the house at the changing of nature, at the passing of time, remembering those we love, sending prayers and blessings their way. Wishing them what we wish for ourselves. Love.

WATER IS FOOD FOR THE SOUL, PURITY FOR OUR MIND AND BODY.

Immersing yourself in a pampering oil bath or in the clear waters of the Sea of Galilee or enjoying the blue hues of the sea and sky and the balcony overlooking the landscape are relaxing, nourishing sources of energy.

THIS IS A REAL VACATION LOCATION;

a soothing place to recharge, energize, and feel content.

A REAL VACATION

is one that enables us

TO GET OUT OF OUR ROUTINE.

Staying in historic, restored rooms is a moving and unusual experience. Knowing that, in its past, the house served in another role stimulates imagination and awakens curiosity. One luxurious unit, with its exceptionally high ceiling, overlooks the lake through the original openings in the basalt wall. Lavish textiles enveloping the room make it exceptionally inviting and indulgent.

设计师与建筑师
DESIGNERS AND ARCHITECTS

Shmulik Aberjel
Page: 125

Rinat Abromovich
Pages: 146, 148–9, 194–7

Maayan Ashkenazy—Reflectura Design
www.reflectura.com
Pages: 62–3, 176–7, 204–5, 228–31, 210–13

Keren Avni
www.kerenavni.com
Pages: 44–5, 114–15

Miri Balbul
www.scullahs.co.il
Pages: 112–13, 136–7

Iris Bar
Pages: 96–9, 156–7, 194–7

Noa Bar-Lev Davidor—Space and Interior Design
www.belonging.co.il
Pages: 11–12, 14–15, 22–3, 66–75, 84–91, 146, 150–1, 158–9, 170–1, 198–9, 224–5, 232–3, 234–7

Kinneret Berkovitch
www.kineretarc.com
Pages: 146, 148–9

Ronit Biton
www.ronitarch.com
Pages: 142–3

Chava Bittermen
Pages: 162–3, 166–7

Merav Cohen Mizrahi
Pages: 20–21, 160–1

Rina Doctor
www.rinadoctor.com
Pages: 6, 26–31, 76–7, 122–3, 218–19

Dov Koren Architects
www.korenarch.com
Pages: 94–5

Tal Eyal
www.taleyal.com
Pages: 100–1

Amit Galor
www.amitgalor.co.il
Pages: 5, 78–9, 154

Anat Gay associated with Jean-Claude Beck
www.gayanat.com
Page: 10

Fred Gurion
Pages: 80–3, 200–1, 226–7

Sara Halkin
Pages: 34–9, 164–5, 284–5

Si Hart
Pages: 94–5

The Heder Architecture
www.theheder.com
Pages: 16–19, 46–9, 190–1

Yaffa Hirsch
Pages: 102–3

Ishai Wilson Architecure
www.ishaiwilsonarchitecture.com
Pages: 272–9

Nurit Kacherginski
www.nuritk.com
Pages: 120–1

Hamutal Katzir
Pages: 140–1

Nurit Kolker
www.nuritkolker.co.il
Pages: 64–5, 180–1, 208–9

Niva Laichter
www.nivalaichter.co.il
Pages: 182–3

Ariela Lavi
www.ariela-lavi.com
Pages: 248–9

Odet Lavy
www.odedlavy.com
Pages: 172–3

Linoy Landau—Home & Photo Styling
www.linoylandau.com
Pages: 50–1, 104–5, 184–5

Ruth Liberty-Shalev
rlshimoor.wixsite.com
Pages: 14–15, 40, 84–91, 150–1, 198–9, 224–5, 234–7

Alon Lotan
www.alonlotan.co.il
Pages: 23–4, 40–1, 150–1, 198–9, 224–5, 234–7

Ludmir Architects
ludmir.co.il
Pages: 140–1

Jonatan Monjack
www.monjack.co.il
Pages: 92–3, 106–7, 110–11, 144–5, 240–1, 252–3

Sharon Miller Maayam—Coastal Living
www.coastal-living.co.il
Pages: 80–3, 200–1, 226–7

Michal Ravet
Pages: 168–9

Rina Noti
notirina.fav.co.il
Pages: 130–3, 188–9

Rimon Architects
www.rimon-arch.com
Pages: 191–3

Michael Ring
www.ringwood.co.il
Pages: 124, 131–3, 188–9

Tal and Roni Ronen
www.habait.com
Pages: 262–71

Nofi Rotem
www.nofidesign.com
Pages: 138–9, 174–5, 216–17

SaaB Architects
www.saab-arc.com
Pages: 242–5

Shaltiel Vartanski Architecture and Interior Design
Pages: 166–7

Vered Shani
www.vny.co.il
Pages: 100–1

Billie Shimoni
www.billieshimoni.co.il
Pages: 162–3

Tami Shimoni
www.nekuda.biz
Pages: 42–3, 126–9, 186–7

Galia Sternberg
Pages: 54–5, 147, 152–3, 250–1

图片版权信息

IMAGE CREDITS

Tav Architects and Designers (Tav Group)
www.tavgroup.com
Pages: 124, 134–5, 286–7

Kinneret Tsidon
www.tsidon.co.il
Pages: 42–3, 126–9, 186–7

Kfir Vax
www.kfirvax.com
Pages: 162–3

Amir Vered
www.amirvered.co.il
Pages: 178–9

Witt Architect
www.witt-arch.com
Pages: 25, 32–3, 280–1

Yana Yarkoni
Pages: 13, 56–61, 116–19

Silvia Yasinovsky Kaplun
www.syk.co.il
Pages: 102–3

Nurit Zeiri
www.nuritzeiri.com
Pages: 246–7

Orit Zilberman
www.tweelingen-design.com
Pages: 44–5, 114–15, 222–3

HEAD PHOTOGRAPHER
Shy Adam
Pages: cover, 5–6, 12–19, 22–3, 26–31, 40–9, 52–69, 71–93, 96–103, 108–9, 114–19, 122, 125–33, 138–43, 146–54, 156–9, 170–1, 174–81, 186–201, 204–13, 216–19, 222–37, 240–5, 250–1, 254–81, 283

ADDITIONAL PHOTOGRAPHERS
Orit Arnon
Pages: 120–1, 220–1, 238–9

Galit Deutsch
Pages: 21–2, 160–1

Yaeli Gabriely
Pages: 124, 134–5, 286–7

Amit Geron
Pages: 32–3

Adi Gilad
Pages: 168–9

Yoav Gurin
Page: 10

Yaffa Hirsch
Pages: 102–3

Yoav Peled
Pages: 92–3, 106–7, 248–9

Gilad Radat
Pages: 8, 25, 33–9, 50–1, 93–5, 104–5, 110–12, 136–7, 144–5, 162–7, 182–5, 202–3, 214–15, 246–7, 252–3, 284–5

ARTWORK
Gadi Dagon
Page: 172

The connection between a

PERSON AND THEIR HOME

is like the connection between

BODY AND SOUL.

Every effort has been made to trace the original source of copyright material contained in this book. The publishers would be pleased to hear from copyright holders to rectify any errors or omissions.

The information and illustrations in this publication have been prepared and supplied by Orly Robinzon. While all reasonable efforts have been made to ensure accuracy, the publishers do not, under any circumstances, accept responsibility for errors, omissions and representations express or implied.

IMAGES has included on its website a page for special notices in relation to this and its other publications. Please visit www.imagespublishing.com